Akashi Seafood,
One of a Kind

The Secrets of Akashi Fish

For a long time, fishery industry has been flourishing in Akashi, which is on the Seto Inland Sea. The variety and flavor of the products fascinates many people. Then, why Akashi fish is so attractive?
Let us explore the world of Akashi fishery products.

The Names of the Local Fishery Products

Akashi-Dai
→ p.22

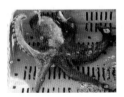

Akashi-Dako
→ p.32

Akashi-Nori
→ p.40

The local fishery products famous for the quality are named after the city, for example, Akashi-Dai (red seabream), Akashi-Dako (octopus) and Akashi-Nori (dried seaweed). It shows how they love these products in the area, and offer them ambassador to the city.

Contents

I love the seascape of Akashi. The sea glistens in the sunshine and the moonlight. Ships and vessels go back and forth. It shows us the power of nature in the day of storm. It provides a home for a wide variety of living and consists of plentiful nutriment. The treasure of the sea makes me more curious about it.

During my service of four years in Akashi as a journalist, I encountered fishermen living with sea, moved by their passion and touched the pride of citizens. This book is filled with stories made by people I admire.

Nami Kanayama (Reporter for the Kobe Shimbun Daily Newspaper)

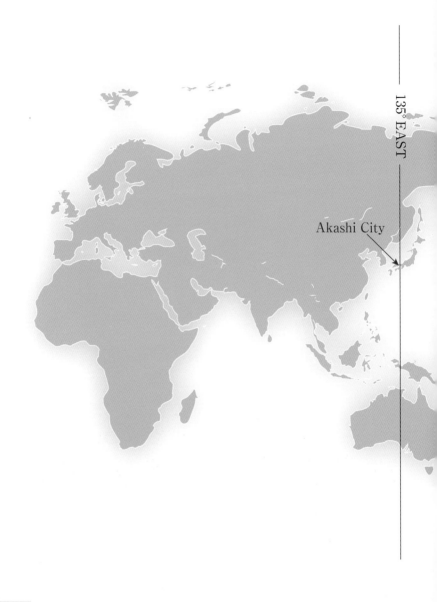

135° EAST

Akashi City

City of Akashi

Akashi is the city located in the southern part of Hyogo Prefecture, exactly on Japan Standard Time meridian of 135° east longitude. The area is 49.42 square km in total. It has a shoreline of 15.9 km facing on Seto Inland Sea (*Harima nada*). You can enjoy the world longest suspension bridge *Akashi Kaikyo Bridge*.

The area has mild climate throughout the year. In spring the thousands of cherry blossom trees are blooming in Akashi Park. We can see the ruins of Akashi Castle, which is registered as a national important cultural asset.

The population is about 300,000. Not only famous for fishery products, it is flourishing agricultural region, for rice, cabbage and strawberries. It takes 3hours by Shinkansen from Tokyo, the largest city of Japan, and also easy to access from Kansai International Airport. In 2017, about 5.6 million tourists visited Akashi and enjoyed these scenic spots and historical structures as well as good foods.

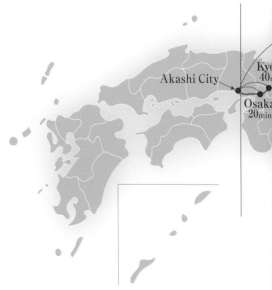

135° EAST

Akashi City

Kyo
40m

Osaka
20min

Tokyo
180min

Shinkansen

Akashi,
the City of Time

There is a 9 hour time difference between The Prime Meridian of Greenwich, UK and Akashi, known as "The City of Time," as Japan Standard Time meridian of 135° east longitude passes through the city. The local people value invisible meridian and time as a symbol of the city.

Local People Admire the Meridian

 You can see various leveling poles and a line on the meridian in the city. People decorate with flowers and learn the history and meaning. The visible line is appreciated as cultural heritage.

Akashi Municipal Planetarium chosen as "No. 1 treasure" by people in the city. The clock tower is built on the meridian. Many visitors enjoy the planetarium. The meridian also runs the platform of *Hitomaru-mae* Station, the nearest station.

The Japanese oldest guide pole on the meridian, donated by teachers of the elementary schools in Akashi in 1910.

Guide poles on the meridian

The post office built exactly on the meridian. They drew lines along with the meridian both on the floor and the ceiling of the inside building, and even in the parking lot.

The guide pole with a dragonfly on the top, a dragonfly is the symbol of *Akitsu-shima*, the ancient name of Japan.

Time is Treasure

Time is equally given to everyone. Standard time allows people to understand time and know time as others do. In Akashi they have many things to remember time everywhere.

The sundial on the façade of Akashi Uenomaru Church, one of the most primitive kinds of clock that tell time by shadow of sunlight.

In Akashi Municipal Planetarium, the educational skits and quiz program provided by Shigosenger and Dr. Blackstar are very popular among kids. Shigosenger (right and left; wear a red and blue uniform) named after *Shigo-sen*: means meridian in Japanese, are the heroes who cover time and space from the attack with poor jokes which pizzle kids cracked by Dr. Blackstar (central; wear a black cloak with violet star on his head).

The sundial in Nakasaki Elementary School, pupils learn the value of time since childhood.

The drum of time in Akashi Park, the Samurai-shaped robot beats the drum on the hour sharp to tell time.

The bell of time in *Gesshoji*-Temple, the bell rings out on June 10th: the Time Day, December 31st: the New Years Eve, and January 17th: the day of Great Hanshin-Awaji Earthquake in 1995.

The certification for passing the meridian, the memorable goods given on the Time Day every year. It is 56th events in 2019 where they distribute various kinds of goods renewed each year; including a towel made by a dye house in local market and a postcard.

The meridian and Akashi Municipal Planetarium are drawn on the manhole cover in Akashi.

Akashi, the City Where People Love Fish

In Akashi facing on Seto Inland Sea, they have inherited a unique living and culture including food and traditional events related to the sea. A variety of seafood live in the good fishing ground of the city. People treat and eat seafood with care so that they learn to be thankful for nature.

The Fertility of Sea

 We have many reasons why the fish in Akashi is so delicious. The first reason is richness of seawater. The wave of inland sea is relatively calm. With Seto Inland Sea in front and mountains behind, the water running from the river has much nutrition.

 The second reason is the depth and current of seawater. The

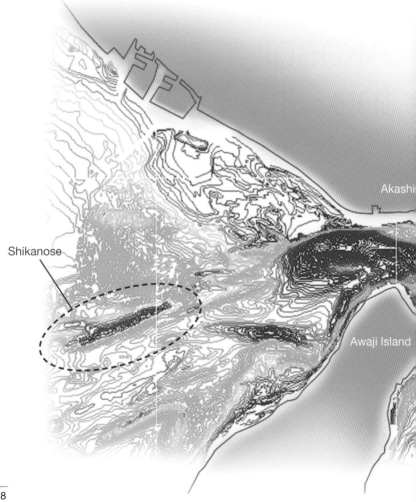

Shikanose

Akashi

Awaji Island

bathymetric chart shows that the channel gets narrow sharply and deepens rapidly around the northern tip of Awaji Island. The tidal current becomes strong enough to give fish good flavor in the area.

The third reason is the shallow called *Shikanose*, shallow of deer in Japanese. That is the perfect place for small fish inhabiting and laying eggs, thanks to gently circulation of water and clean sand bottom of the sea. School of small fish lures other kinds of fish, including red seabream and octopus. The whole marine food chain may keep these abundant varieties of fish.

Local fishermen are able to get various kinds of fish because of richness of the sea. In addition, they put focus on conserving marine resources in Akashi.

Final reason is the closeness between the sea and the fishery port. Fishermen deliver and sell by auction fish fresh from sea. The scenes of the fishery auction are familiar to local people. The fishermen improve the technique to deliver their fishery products in the best condition to consumers.

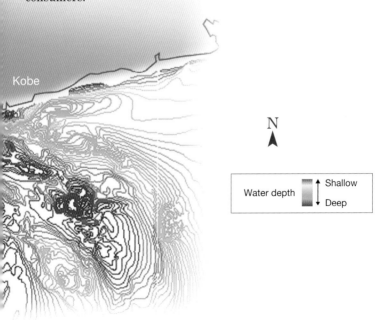

N

Water depth — Shallow / Deep

The Fish Masters

A tolling tells the beginning of the auction. Wholesalers show their products, and intermediate wholesales (*Nakaoroshi-gyosha*) put a bid to the fishery products. The buyer who made the highest bid obtain the products. The auction is called "*Hiru-ami*" (literally meaning: daytime's catch) as it is held around noon in Akashi. The various kinds and amount of fish come one after another, the sellers dashingly call out, the buyers put a bid with hand sign and then, the deals close so quickly. The vibrant fishery auction full of salty air, spray of water and serious eyes of sellers and bidders, this is one of the most-beloved scenes in Akashi.

You won't find such an auction dealing in fish fresh from sea anywhere. Akashiura Fishery Cooperative Association makes it possible with the pool running seawater for 24 hours.

The auction in Akashiura Fishery Cooperative Association. Visitors are admitted after making an application.

The wholesale market also attracts many buyers.

Akashi-Dai

Red seabream has been considered the "king of fish" in Japan since early times, because of the taste, shape and color. Especially Akashi Red seabream (Akashi-Dai) is well known as the one of the best brand-named, wild-caught and the finest white-fleshed fish.

鯛

鯛

23

Main Features of Akashi-Dai

 People admire the pinky colored red seabream with seasonal pet names: in spring they call it "*Sakura-Dai*", cherry blossom red seabream, named after beautiful pinky color of female fish laying eggs. In autumn they call it "*Momiji-Dai*", turning leaf red seabream with slightly reddish color with plentiful feeding. The fishing season starts in April until December. Only 20% of red seabream distributed in Japan is wild-caught, the other 80% is farmed. Wild caught Akashi-Dai has distinctive appearances.

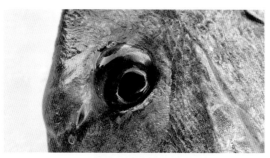

A blue line on their eyes, just like eye shadow. Nose holes clearly separate into two. (e.g. farmed red seabream has only one nose hole.)

Shiny clear color of skin, blue spots in the back.

Deep red tail fin, pectoral fins and trail fin are upturned.

Naruto-bone, the characteristics back bone with nodes. Small bone in the shape of fish in pectoral fin joint is called *Tai-no-Tai*: tiny red seabream inside red seabream. Someone carries it as good-luck charm to have comfortable life.

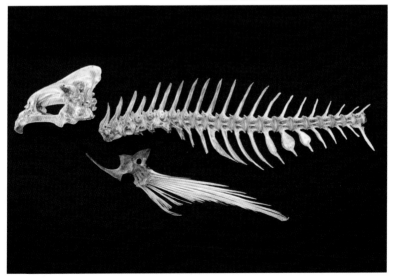

Fishing Methods

The most common method of red seabream fishing in Akashi is what called "*Gochi-ami ryo*," that is *Gochi*-net fishing method. *Gochi* has origin in Buddhist terminology which means wisdom beyond the self. It implies that only wise fishermen could handle the method. They locate the point to drop a net according to the course recorded by GPS: Global Positioning System and the points caught fish in previous day, and tide at the moment. Other methods are fishing rods and trawling

Fishermen drop and draw up a net with approximately 400m-length rope by a winch. It takes 15minutes for one routine, they repeat it 20 times a day.

Pinkly red seabream in the net. The results depend a great deal on their gut feeling, experience, strength and smooth operation between crewmembers.

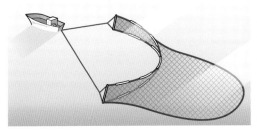

Gochi-ami ryo, *Gochi*-net fishing method. They drop an ellipse shaped net with rope each end, drive a fishing boat in a wiggly way like U figure to make the net like a bag. Using strength of against tide, they catch red seabream into the net which width is approximately 30m.

Traditional Methods to Ensure Freshness

Akashi-Dai has been recognized as Japan's No.1. It attracts customers not only in large cities in Kansai area, including Kyoto, Osaka and Kobe, also in Tokyo, the Japan's largest market, and even in other countries in Southeast Asia. The technical capability to ensure freshness supports this reputation. These traditional techniques are called *Akashiura jime*, Akashiura technique, established and carried by people involved fishery industry who are versed in the sea, and characteristics and structures of fish.

The treatment provided on the ship is deflation. Crewmembers put a caught red seabream upside down, and insert a needle through the anus. Red seabream is caught in 100m depth of sea and the sudden change of water pressure makes its air bladder swell so that it cannot swim straight. To keep the freshness, they make a small hole in red seabream's air bladder without killing.

As soon as landing, red seabream is put into a dark tank in Akashiura Fishery Cooperative Association. It is called *Ike-goshi*, that is, they keep the fish over night in the dark tank, to make them calm down and remove the stress. In the same time, it makes them vomit things inside their stomach and avoid spreading smell into the flesh.

Alive red seabream bought by bidders is taken to next procedure, *Ike-jime*. *Ike-jime* is a Japanese original technique the keeps fish fresh by

Deflation, making a small hole in the air bladder without damage to other internal organs. It needs careful and proficient skills.

Ike-goshi, spending over night in the dark tank. The flesh is coarse just after caught. After making them calm, it will ensure freshness and slowly progress postmortem rigidity when it would be killed.

immediate death.

Next, they give *Shime-bocho*. The associate cut off the backbone inserted the knife through the operculum to exsanguinate.

Then, they give *Shinkei-nuki*, taking out nerves. They break up the spinal nerves by threading upper side of backbone with wire. It makes them postpone preceding postmortem rigidity to keep freshness by hindering transmission of the information of the death from nerves to each cell.

At the same time, they also put focus on the appearance. Fish is placed on the dish with its head on the left, meaning the left side of body will be upper side. Every treatment is designed to avoid scratch on that side. For example, *Shinkei-nuki* is executed by pierced with wire from the crack of scale neat the tail fin.

Another persistence is shipment temperature. They set particular temperature best for shipping, based on the data provided by the researchers of professional organization.

They deal taste and appearance in the same time. Every treatment is provided carefully and speedily. The brand of Akashi-Dai is supported by their professional techniques.

Ike-jime, killing instantly by a hook. Without fatal hit, red seabream run wild and it will cause bruise.

Shime-bocho, cutting off the backbone with knife. It avoids flowing blood inside body, it leads lose quality.

Shinkei-nuki, taking out nerves with wire. The technique is provided to postpone postmortem rigidity.

Cooking **Akashi-Dai**

Red seabream attracts many people with its unique flavor and taste. They have various way of cooking, such as sashimi, grill, steam, roast, parboil, and soup stock. It is used any kind of dishes; Japanese, Italian, French, Chinese or others. It contains of high protein and highly nutritious as well as low fat. With its gorgeous shape, it has been used as an auspicious food.

Akashi-Dai is not soft just after caught. After postmortem rigidity finished, it becomes tender. Aged red seabream for a few days under appropriate temperature is good for sushi, as protein decomposes into umami elements. Skilled chefs are fascinated with Akashi-Dai to make their effort to cook under the best condition for each dish.

Shiny Akashi-Dai before cooking. The only certificated one has the trademark on the pectoral fin.

Shio-yaki, salty roast of the part of pectoral fins with plenty of fat.

鯛

Ara-daki, after parboil, boiled bone parts with sake, soy source, sugar and soup stock. No one can resist the glossy source covering the fish.

Sushi garnished with grated daikon, finely sliced welsh onion and soy source with citrus juice.

Yaki-dai, grilled red seabream is served on auspicious occasion: a celebratory meal and New Year's plates. For keeping its beautiful shape, chefs grill it one bye one; fix up the fins with salt and a bark of *hinoki*: Japanese cypress, truss with two skewers and adjust right heat level for each.

Salt and broil red seabream, broiled encrusted with salt and egg albumen. The breaking presentation with a hammer will warm up a party.

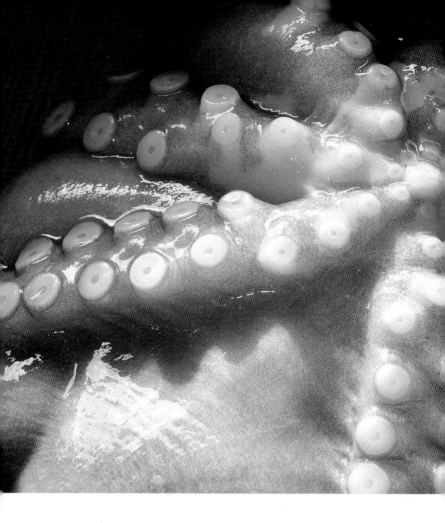

Akashi-Dako

The octopus is a soft-bodied, eight-limbed mollusc living in the sea.
Akashi-Dako, a kind of "*madako*," which is most popular in Japan.
It comes its season in summer, with highly nutritious and taste. Local
people love Akashi-Dako, which is a highly recognized brand octopus in
the country, and makes the city even more famous.

蛸

蛸

Main Features of Akashi-Dako

A common octopus (*madako*) has relatively short lives. It lives as much as 12 to 24 months. It is vulnerable to cold, and dies in seawater at 5-6°C. Off the coast of Akashi, its spawning season starts in middle of August, and eggs hatch by the year end. After surviving cold of winter, octopus grows up twice in 15 days in warm spring. As octopus eats shrimps, crabs and shellfish, it contains plenty of taurine. The longer you chew it, the better you can enjoy the flavor.

Thousand tons of octopuses are landing annually, mainly common octopus, the others are gold-spot octopus and long arm octopus.

Akashi-Dako keep its footing against strong tide near the strait, so that it has stubby arms and its flesh is firm and muscular.

They have 200-240 sukers for each arm. Female's suckers are aligned regularly and male's are out of order.

Traditional methods to capture octopuses include pots and fishing, according to its clinging characteristic.

In the shopping street called "Uontana" some Akashi-Dako try to escape as walking.

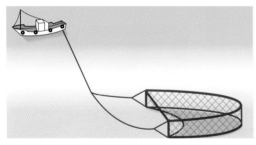

Fishermen drawing up the net on the trawling boat. In a shallow sea, they move the boat roundly to drag a rope on bottom of the sea and lead the octopuses into net. In a deep sea, they trawl the net against to tide with dragging weights on bottom of the sea, and make octopuses follow into end of the net.

Fishing Methods

They have three methods to capture octopus. The most popular method is *Sokobiki:* trawling, pulling the fishing net through the bottom of the sea behind a boat. The second traditional method includes pots called *Tako-tsubo.* They make use of its unique habit of octopus of hiding in gaps of rocks. With the method, they capture octopus in good shape without deface. The final method with lure allows them to capture fresh octopus.

In those cases, fishermen compete with octopus like never-ending game of wits.

A fisherman drawing up the pots sunk into the sea, which are fixed with rope, by using a winch. He adjusts the speed of the winch to keep octopus inside the pots.

Octopus coming out from a pot. Less spilled water during drawing up and much spray water during turning is a sign to catch octopus, appear about 30% of probability.

Handmade lures made of bamboo and cotton wound by colorful plastic tape with large fishing hook on each side, what is called candy.

A fisherman throwing lures. He repeats to cast and draw up regularly. The experts could handle with 12 lures in the same time.

Cuisine of **Akashi-Dako**

Octopus is prepared with various ways. It is used for Japanese-style food, Italian and Spanish as well. Japanese snacks with octopus ingredient, such as rice cakes are very popular for souvenirs. *Akashi-tamago-yaki*, the most famous local specialty which is a ball-shaped pancake grilled on the copperplate with round hollow, has small piece of octopus inside. On the outside, it looks like takoyaki, but has a softer and more eggy texture than takoyaki.

Sashimi, people enjoy the freshness unique to the region.

Tempura, standard dish of octopus.

Simmered octopus with small taros in a mixture of sugar and soy source.

Vinegared octopus with cucumbers. Simple taste.

Full-course dishes with octopus are served at restaurant.

蛸

Hoshi-dako, a preservation of octopus by drying under the sun in summer. Removed the guts and ink, stretch the body and arms with bamboo sticks and hang with a wire or a string to dry.

Octopuses boiled after kneaded with salt, turn into russet.

Akashi-Nori

Akashi-Nori has luster in deep black color, distinctive aroma, and rich nutrition. It smoothly melts in your mouth. Many people are devotedly engaged in the production in the sea and on the seashore. Each dried sheet is filled with the blessings of nature and the spirit of fishermen and producers.

海苔

Growing **Nori Seaweed**

They have no time-off in the cultivation, though the growing period of nori seaweed is from December to May. Production and processing of nori is well developed. Farmers plant filamentous spores inside oyster shells in summer. They seed carpospores produced by filamentous spores onto nets in autumn and grow them in the sea. They blend a few species of microscopic seeds according to the characteristics and conditions. Nori seaweed grows up with plentiful nutrition in the sea.

Filamentous spores; preserved in a flask. They are working on breeding improvement in a research institution.

Conchocelis filamentous spores make oyster shells dissolve and grow in black. In autumn, they provide huge amount of seeds.

Farmers place nets into seawater that filled with spores and plant on the net by using a large wheel.

Farmers suspending nets operating a one-person small boat, rocking on wave and hanging out of the boat.

Cultivation

From late autumn to spring in Akashi, you can see square and black nori seaweed nets in the sea. The method called *Uki-nagashi-shiki* is popular there. Utilizing anchors, buoys and ropes, farmers cultivate nori seaweed even in deep water and strong current. Near the end of November, when seawater temperature becomes about 18°C, they start to suspend the net attached nori seaweed seeds on the sea surface. The plants grow rapidly; they grow up approximately 20cm within 10days. Farmers pick up nori seaweed by passing under the lifted net, which is called *Moguri-sen*. To cultivate good nori seaweed, they pay great attention to temperature, weather and humidity, put their experience and knowledge, and always devote yourself to the work.

Moguri-sen boats lift nets and go under them. The net drip down seawater and crewmember pick nori seaweed.

It is winter feature in Akashi, many nori seaweed nets on the sea and *Moguri-sen* boats come and go.

Production of Akashi-Nori

Just after *Moguri-sen* boat returns, crewmembers connect a hose to tank to send just-picked nori seaweed. In factories, nori seaweed is minced, formed into plate-like and dried. Finished products are brought to quality check. Examiners divide into about 100 grades depending on its color, gloss and shape. After that, wholesalers obtain at auction. They productize them through these processes, including secondly drying, toasting, seasoning, cutting and packing.

Tanks filled with picked nori seaweed on the boat.

They are connecting a hose with tanks on the boats and tanks on the land to send to tanks.

Many tanks and factories placed side by side are called as like nori industrial park. Experts handle production procedure, including adjusting the amount of water to be required when minced and drying time depends on the condition of nori seaweed.

After density adjustment, nori seaweed is put on a mat to remover water as like papermaking, suction and drying. Nowadays with highly automated machines spread in the industory, the processes are accomplished by the machine.

Examiners evaluating the grade with serious looks.

At auction, wholesales estimate the value by look, taste and smell.

Seasoning process line in wholesalers. Different company has different recipe for mixing, drying and cutting process for putting into production.

Culinary **Akashi-Nori and Products**

Nori seaweed has plenty nutrition, for example, vitamin, minerals and protein. The high-end products is called *Ichi-ban-tsumi*, the first of the year's harvest are delicious and nutritious. Many products of nori seaweed are popular in the market. Seasonings with nori flavor, noodle and snacks using powdery nori satisfy customers as it is much easier to digest.

Gift boxes of 5 brands of nori made by fishery cooperative associations in Akashi. It also pleases customers to compare the flavors between them.

Dressings, soy source and ramen flavored Akashi-Nori with distinctive aroma.

Cosmetics and facial soaps consist of Porphyran, soluble dietary fiber, which is highly contained in certain seaweed to protect cell from ultraviolet rays and dehydration. Nori seaweed has unlimited capability.

海苔

Gunkan-maki, sushi with sea urchin or salmon roe on the top

Nori-maki, sushi roll with cucumber, Japanese omelets and boiled whitespotted conger inside. In the day of *Setsubun*, the last day of winter in lunar calendar, local people have s a custom to eat whole size sushi roll without cutting.

Hand-rolled sushi, put vinegar-flavored rice and favorite fillings on hand size nori, and just wrap it to make it.

Boiled whitespotted conger, one of the famous local foods in Akashi, with nori. Suitable combination for a snack.

Our Wish for Abundant Marine Resource

All people concerned with the sea in Akashi wish for its fertility and sustainability forever. Sharing thankfulness for nature, people in fishery industry in the city have been working on initiatives together with citizens for conserving environment, passing the wisdom on to next generation, and developing their city brand.

Initiatives for Sustainable Sea Environment

Ecosystems in sea change incessantly. In 1990s, a lack of sea nutritive salts mainly consisting of nitrogen and phosphorus frequently caused nori seaweed color loss. Then, people put focus on activities for cleaning up seawater pollution in 1970s to 1990s, and now, their focus shifted to biodiversity protection. They make steady efforts to prevent from overfishing and for conservation of recourses.

Fishermen throwing pots for octopus to stay and laid eggs in the sea. This method has been continued since 1966.

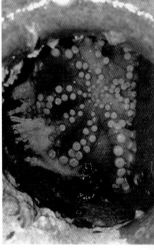

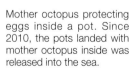

Mother octopus protecting eggs inside a pot. Since 2010, the pots landed with mother octopus inside was released into the sea.

Kaibori, a traditional maintenance method of agricultural reservoir, helps nutritional cyclical process from field to sea. Farmers, fishers, students and children work together to protect natural environment.

Sampling fish juvenile. For sea resource conservation, the opening day of fishing season has been decided based on the distribution and assumption of growth.

Brand Identity

 They gave the name of the city to red seabream, octopus and nori of their area. Thus, the name of Akashi has been spread widely to establish the strong brand. They work on designing attractive experience for customers. Through enjoying the event, participants find value, know fishermen's efforts and share their dedication. All people in the city make contribution to improvement of the brand.

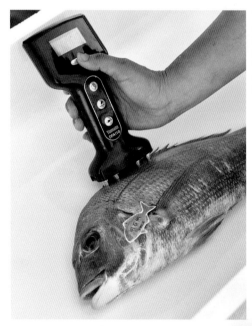

The fish analyzer, the equipment scored freshness and plumpness of fish. The fish shaped trademark on pectoral fin is the proof of certificated Akashi-Dai, weigh 800g or more.

The highest grade of certificated Akashi-Dai shipped in special black box.

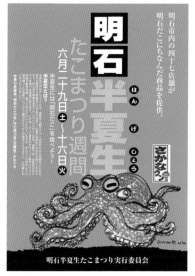

Poster advertising for educational program about Akashi-Dako held in every July, and about Akashi-Nori in every January.

They have even a character of octopus as community symbol for promoting Akashi tourism. *PaPa Tako*, Daddy Octopus, is popular among kids.

Kids enjoying the octopus shaped slide in Matsue Park.

Akashiura FCA
Hiroaki Ebisumoto

We're very proud that our products have good reputation throughout Japan including Tokyo. All associates work hard and improve their skills for delivering better products for customers. I devote myself to sending messages to get people to know more about our efforts to improve the quality and its taste.

Hayashizaki FCA
Masao Tanuma

Akashi has been recognized fertile fishery region since early times. Recently, we fishermen have an initiative for sustainable sea resources with support of citizens. It is one of our accomplishment with organizations of Hyogo prefecture to lead the revision of Act on Special Measures concerning the Conservation of the Environment of the Seto Inland Sea in 2015. Sea resources is our treasure for all generations, and our mission is not only to prosper in the fishery industry also to conserve the nature.

Eigashima FCA
Mikiya Hashimoto

We fishermen are very proud of quality of our products as one of the World's best. For sustainability of the industry, we regulate fishing in this region with a policy such of fishing quotas. Akashi is the city of sea and fishery. I'm grateful if people know more about this city of marine richness through our products.

Higashi Futami FCA
Yoshio Ohnishi

It is essential that five FCAs work together as one team. Our educational and PR programs, and preservation initiatives give the city much more than respective associations could. We're very happy to see visitors to Akashi enjoying one of the best fish in Japan and tasting local good products in the region.

Nishi Futami FCA
Akira Yamamoto

My responsibility is to give information on our fishing method, dedication, processes of delivering to consumers, conditions of the sea and any other aspects concerned to our business. Sea resource has unlimited potential. We would appreciate if more people would get more familiar with this city through our products.

Messages from Presidents of five Fishery Cooperative Associations in Akashi

Federation of Akashi-city Fisheries Association work on sustainability of sea resource, as well as give much information to local people for better understanding of their business. They share pride, skills, knowledge and affection for the city. The fishery products in Akashi fascinate local people as they develop the idea and passion of the people involved in fishery industry in the city.

Akashi Fish, our Pride and Joy

Ako
(Redspotted grouper)

Aburame
(Greenling)

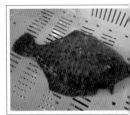

Amagarei
(Marbled flounder)

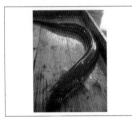

Anago
(Whitespotted conger)

Bakeshita
(Three-lined tongue sole)

Chinu
(Blackhead seabream)

Gashira
(Marbled rockfish)

Hamo
(Daggertooth pike conger)

Hariika
(Golden cuttlefish)

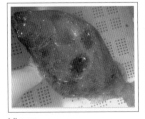

Hirame
(Bastard halibut)

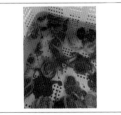

Iidako
(Gold-spot octopus)

Ikanago
(Pacific sandlance)

local names in Akashi
(English names cited from https://www.fishbase.de/search.php and https://www.sealifebase.se/search.php)

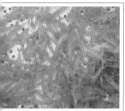

awatsuebi
(outhern rough shrimp)

Maruhage
(Threadsail filefish)

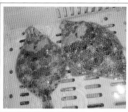

Meitagarei
(Ridged-eye flounder)

miika
(apanese bobtail squid)

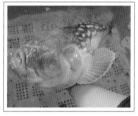

Okobo
(Spotcheck stargazer)

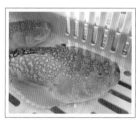

Okobo
(Japanese stargazer)

koze
(corpionfish)

Sawara
(Japanese spanish mackerel)

Suzuki
(Japanese seabass)

chiuo
(argehead hairtail)

Tamori
(Broadbanded velvetchin)

Tsubasu
(Japanese amberjack)

Author

THE KOBE SHIMBUN DAILY NEWSPAPER

The Kobe Shimbun Daily Newspaper is one of the major local newspapers in Japan founded in February 1898. It has the main office in Kobe and a great many readers in the area. As of first half of 2019, its daily average circulation was 485,599 for morning paper and 148,559 for evening paper, according to Japan Audit Bureau of Circulations. When the Great Hanshin-Awaji Earthquake destroyed the main building and ruined the publishing systems of the newspaper company in 1995, they devoted themselves to writing articles on the disaster and publishing their papers every day for people who needed information, with support of their friendship company, the Kyoto Shimbun. In November 2012, it launched the online edition, "Kobe Shimbun NEXT." The Sun Television, The Radio Kansai and The Daily Sports Daily Newspaper are the affiliated companies.

Akashi Seafood,
One of a Kind

First edition 1st printing March 5, 2020

Author THE KOBE SHIMBUN DAILY NEWSPAPER
English translation by Naoko Ishiwata

Publisher Yoshiaki Kawasaki
Federation of Akashi Seafood Wholesalers Cooperative Association
Publishing office Pencom Co., Ltd.
2-20 Hitomarucho, Akashi-shi Hyogo Prefecture 673-0877
http://pencom.co.jp
Release Impress Corporation
105, 1 - chome, Kanda Jimbocho, Chiyoda ku Tokyo 101 - 0051

English supervised by Toshimi Aishima,Ph.D
Publishing Yukimi Masuda, Takashi Aoki
Design Yuko Isawa

For inquiries about marine products,
mailto:qqzv7ar9k@bell.ocn.ne.jp

Akashi Seafood,
One of a Kind

2020年3月5日　第1刷発行

著　者　神戸新聞社
翻訳者　石綿奈穂子
発行者　川﨑喜昭（明石海産卸売協同組合）
発　行　株式会社ペンコム
　　　　〒673-0877　兵庫県明石市人丸町2-20　https://pencom.co.jp/
発　売　株式会社インプレス
　　　　〒101-0051　東京都千代田区神田神保町一丁目105番地

●本の内容に関するお問い合わせ先
　株式会社ペンコム　TEL：078-914-0391 FAX：078-959-8033　office@pencom.co.jp
●乱丁本・落丁本などのお問い合わせ先
　TEL：03-6837-5016　FAX：03-6837-5023　service@impress.co.jp
　（受付時間／ 10:00-12:00、13:00-17:30 土日、祝日を除く）
　※古書店で購入されたものについてはお取り替えできません。
●書店／販売店様のご注文窓口
　株式会社インプレス受注センター
　TEL：048-449-8040　FAX：048-449-8041
　株式会社インプレス出版営業部　TEL：03-6837-4635

英文監修：相島淑美（関西学院大学 講師）
編集：増田幸美、青木 崇
装丁：井澤裕子
印刷・製本　株式会社シナノパブリッシングプレス

SuperThanks

Japanese food Sukerokugozen, Kiraku
Location AKASHI PARK

Thank you everyone who cooperated with us, and Bounties of Akashi
–Akashi-Dai, Akashi-Dako, Akashi-Nori.